COLORING HISTORIC THEATRES

MICHIGAN THEATER
AND
STATE THEATRE

a coloring book for adults

Text and illustrations by
ESCOTT O. NORTON

presented by

 MARQUEE
ARTS

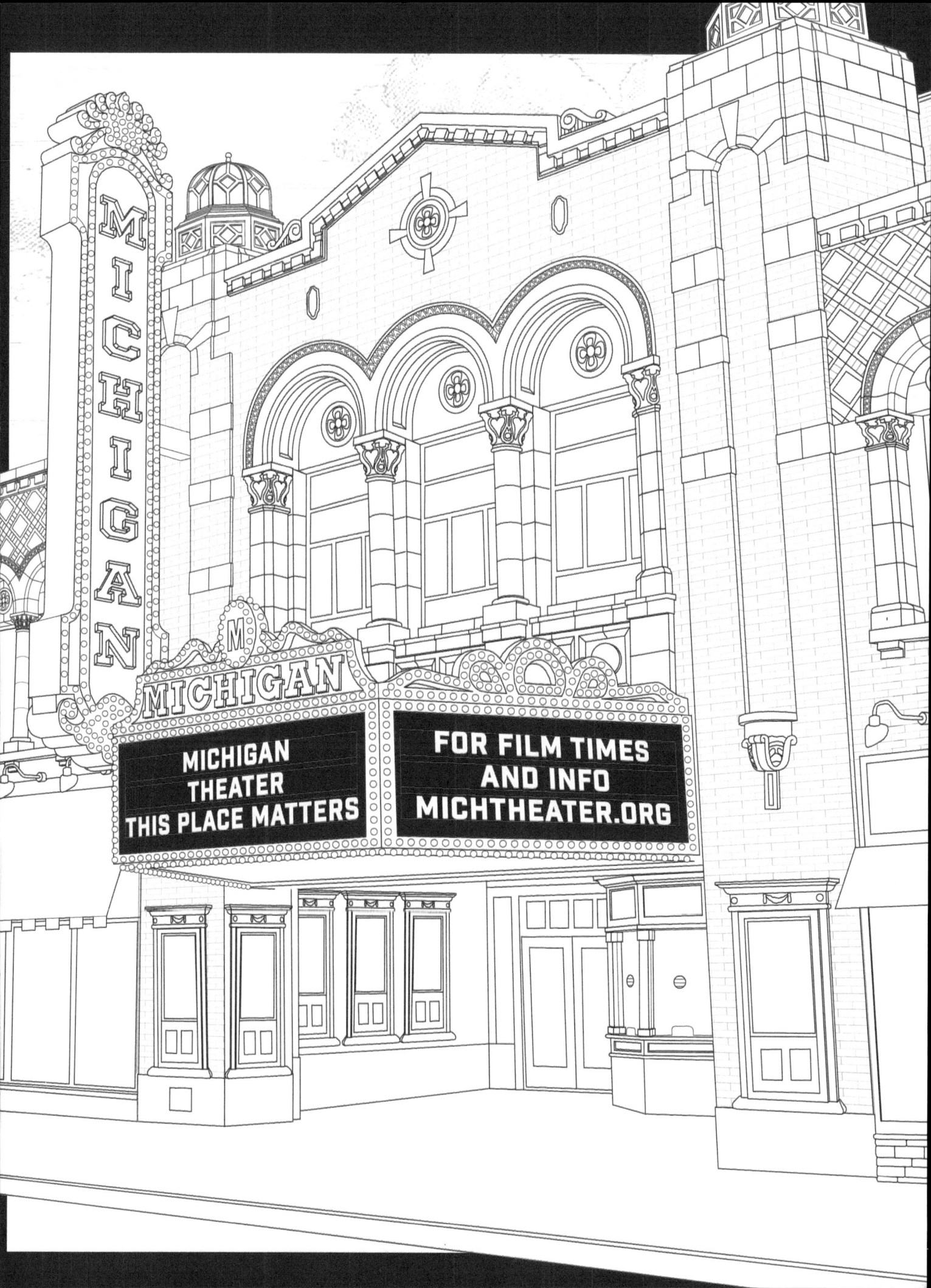

MICHIGAN

MICHIGAN

MICHIGAN
THEATER
THIS PLACE MATTERS

FOR FILM TIMES
AND INFO
MICHTHEATER.ORG

Michigan Theater Exterior

The Michigan Theater opened on January 5, 1928. The opening-night program featured Ida Mae Chadwick and Her Dizzy Blondes in a stage production called From Rags to Riches, followed by the feature film, A Hero for a Night.

Michigan Theater Marquees

The Michigan Theater has had a total of four marquees over the years. The marquee on opening night was a traditional rectangular design, typical of the era. It used a milk glass letter system so it appeared that the letters were white on a black background. The spacing between letters was blacked out using pieces of painted metal.

In an effort to "modernize," the original marquee was replaced in the 1940s with an art deco marquee with neon lights. This was also the first marquee to use colored letters on a backlit white background.

The third marquee was installed in 1956. It was a simpler design featuring a much larger letterboard; it was triangular with angled sides, allowing for easy viewing from passing cars.

The fourth and current marquee (page 3) was completed in 2002. Stylistically it harkens back to the original rectangular marquee with a contemporary version of the milk glass letters, but is trapezoidal in shape.

Activity:
Use the blank marquees in the illustrations to add your own favorite movie or message. Can you match the look?

MICHIGAN THEATER
ANN ARBOR

Michigan Theater Architecture

In addition to its monumental status in Ann Arbor, the Michigan Theater is also an outstanding historical preservation success story. Threatened with demolition and conversion into a food court in 1979, it was saved by the combined efforts of concerned citizens.

During the "roaring twenties," an era of high style and lofty ambition, Detroit architect Maurice Finkel designed a grand silent film exhibition theater appropriate for a town with a world-class university.

Ann Arbor was a burgeoning cultural hub and eager to add a venue suited for the popular entertainment of the day. In addition to silent film, the community flocked to organ concerts and orchestra arrangements, thrilling vaudeville acts and national celebrities on tour.

Born in Romania in 1888, Maurice Finkel immigrated to the United States with his family when he was a child. First discovering his love of theaters as a Yiddish theater actor, he moved to Detroit, Michigan after earning a degree in Architecture in New York. He designed more than 200 buildings, including grand movie palaces like those he'd admired and performed in in New York.

MICHIGAN THEATER ANN ARBOR

Grand Foyer Ceiling Detail

The Michigan Theater lobby is highly detailed, consisting of a series of repeating indented octagons and squares. These indentations create what is called a "coffered" ceiling. Often these coffer details are cast in plaster as one piece, making this highly detailed ceiling easier to mass produce and install.

MICHIGAN THEATER
ANN ARBOR

Michigan Theater Restoration

Restoration formally began on Monday, May 5th, 1986. The walls of the auditorium plaster treatments were restored and repainted to replicate the original 1928 color palette. New carpeting was installed to match the original design created.

When patrons returned to the Michigan Theater on Saturday, September 20th, 1986, they were greeted by the original gold, blue, and red tones of the historic Grand Foyer. Beneath their feet, a plush reproduction of the 1928 carpet with its original pattern.

The second phase of this initial restoration campaign began in 1998 and was completed in 2002, featuring new and improved seating throughout the main floor.

MICHIGAN THEATER ANN ARBOR

Michigan Theater Plaster Treatments

The decorative plaster treatments on the auditorium's side walls and organ grills were demolished in the 1956 modernization. Fortunately, a few pieces of the original plasterwork had survived as rubble in the walls. These fragments were used to make new molds which were used in the restoration project. Paint scrapings showed that the original palette for the Michigan Theater was richly multicolored, with the plaster treatments predominantly red and gold. These were painted over multiple times from the 1930s into the 1950s.

Along with the original drawings and some photographs of the auditorium captured throughout the decades, the architects were able to prepare accurate plans for complete restoration - including niches, exit arches, organ chamber arches, and grids.

MICHIGAN THEATER ANN ARBOR

Introducing Marquee Arts

The Michigan and State Theaters are preserved, programmed, and operated by a dedicated team, now known as Marquee Arts. Thanks to the generous support of our community, our iconic marquees continue to shine brightly over downtown Ann Arbor and serve as a beacon to arts lovers everywhere.

Thank You To Our Sponsors & Donors

with special thanks to

The Silent Classics Film Fund

Silent film programs are made possible by a gift from the estate of Bud Bates, who for over 50 years, was one of the first and most popular volunteer organists at the Michigan Theater. His musical gifts and dedication helped create and sustain the renaissance of this very special Barton Organ.

Dick & Norma Sarns Fund for the Michigan Theater Foundation,
an endowed fund at the Ann Arbor Area Community Foundation.

And to our generous Board members

Board Members 2023-24:

Michigan Theater Fountain

This bronze drinking fountain was discovered in a storage room in the basement of the theater.

The fountain is the same as New York's famous Roxy Theater at West 50th Street; this duplicate is original to the Michigan Theater.

It is believed that the fountain was manufactured by either J.L. Mott or J.W. Fiske, both of New York, who were renowned in the 19th Century for their foundries.

This fountain is no longer operational, but has been beautifully restored and installed permanently as a center fixture in the Grand Foyer.

MICHIGAN THEATER
ANN ARBOR

Michigan Theater Proscenium

When the Michigan Theater opened, it had a traditional stage with orchestra pit in front of the proscenium. The "proscenium" is the frame that separates the stage house from the auditorium. As was common in early cinema-houses, there was an asbestos fire curtain that could be lowered in case of fire to separate the stage house from the auditorium.

The orchestra pit is now covered by a "thrust" stage, meaning a stage that extends out in front of the proscenium.

MICHIGAN THEATER ANN ARBOR

The Barton Organ

The Michigan Theater is proud to be home to its original Barton Theatre Pipe Organ. Installed in the Main Auditorium in 1928, it is still played today before movies and silent films, lectures, and special events.

In the 1920s, every movie palace had a fabulous pipe organ to accompany silent films. Over 7,000 were built, but only a few hundred of these instruments survive, and less than forty remain in their original homes. Fortunately for the Michigan Theater, our Barton Organ stayed in place, even if idle, until 1971, when a team of dedicated organists led by Henry Aldridge revived and restored this musical gem.

Starting in 1972, Aldridge, Rupert Otto, and Newton "Bud" Bates were among those dedicated organists who returned the Barton to a regular performance schedule.

MICHIGAN THEATER
ANN ARBOR

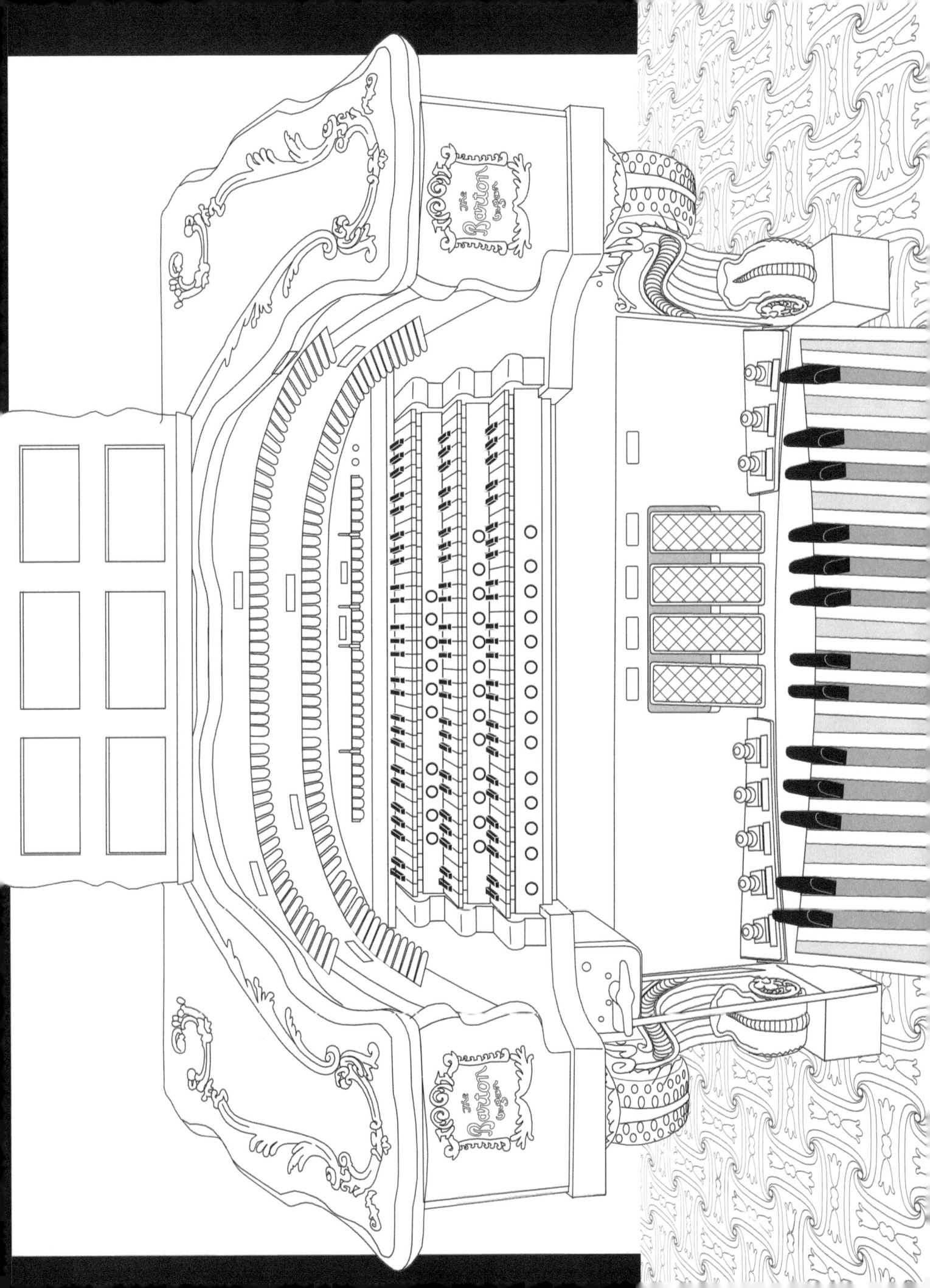

Michigan Theater Organ Grilles

On either side of the auditorium is a group of five archways. These serve as the organ grilles. Behind the grilles is a room called the organ chamber. The organ console visible on stage is just the "tip of the iceberg;" most of the organ is out of sight, including a blower mechanism, sound effects, and most importantly, the organ pipes themselves.

Photos courtesy of David Hufford

MICHIGAN THEATER ANN ARBOR

The State Theatre

The State Theatre opened on March 18, 1942, with a screening of a Paramount musical starring Dorothy Lamour, William Holden, Eddie Bracken, and the Jimmy Dorsey Orchestra. A newsreel and a Warner cartoon - Rhapsody in Rivets - were also on the program.

W. S. Butterfield Theaters began construction on the new theater in the early 1940s. When the United States entered World War II in December 1941, Butterfield theaters made great efforts to assure the community that no materials used in the construction had been taken from the war effort.

The State Theatre's design is heavily influenced by the Art Deco movement. The exterior of the State is in near original condition and is reminiscent of an earlier example of Art Deco theater design - Radio City Music Hall.

STATE THEATRE
ANN ARBOR

State Theatre Renovation

The State Theatre was designed by Michigan architect C. Howard Crane. Crane is the same designer as the Fox Theater and Orchestra Hall in Detroit, as well as over 200 other theaters around the world!

When the State first opened, it seated 1,900 and was slightly bigger than the Michigan Theater. Its interior was simple but elegant, featuring rose and green tones with soft lighting.

While the exterior marquee had remained mostly in its original Art Deco design, the interior had experienced significant alterations over the decades. In the 1970s, it was divided into two theaters, then split again into four theaters in the 1980s when the main floor was converted into retail space, with the theaters remaining in what used to be the balcony.

In 2017, the State Theatre underwent a historically sensitive renovation by Marquee Arts. The space was crafted into four fully accessible movie theaters, featuring improved sight lines, contemporary projection and sound equipment, and a new escalator and elevator.

State Theatre Tile Work
The tiles featured in the next illustration were crafted by local artist Nawal Motawi of Motawi Tileworks.

STATE THEATRE ANN ARBOR

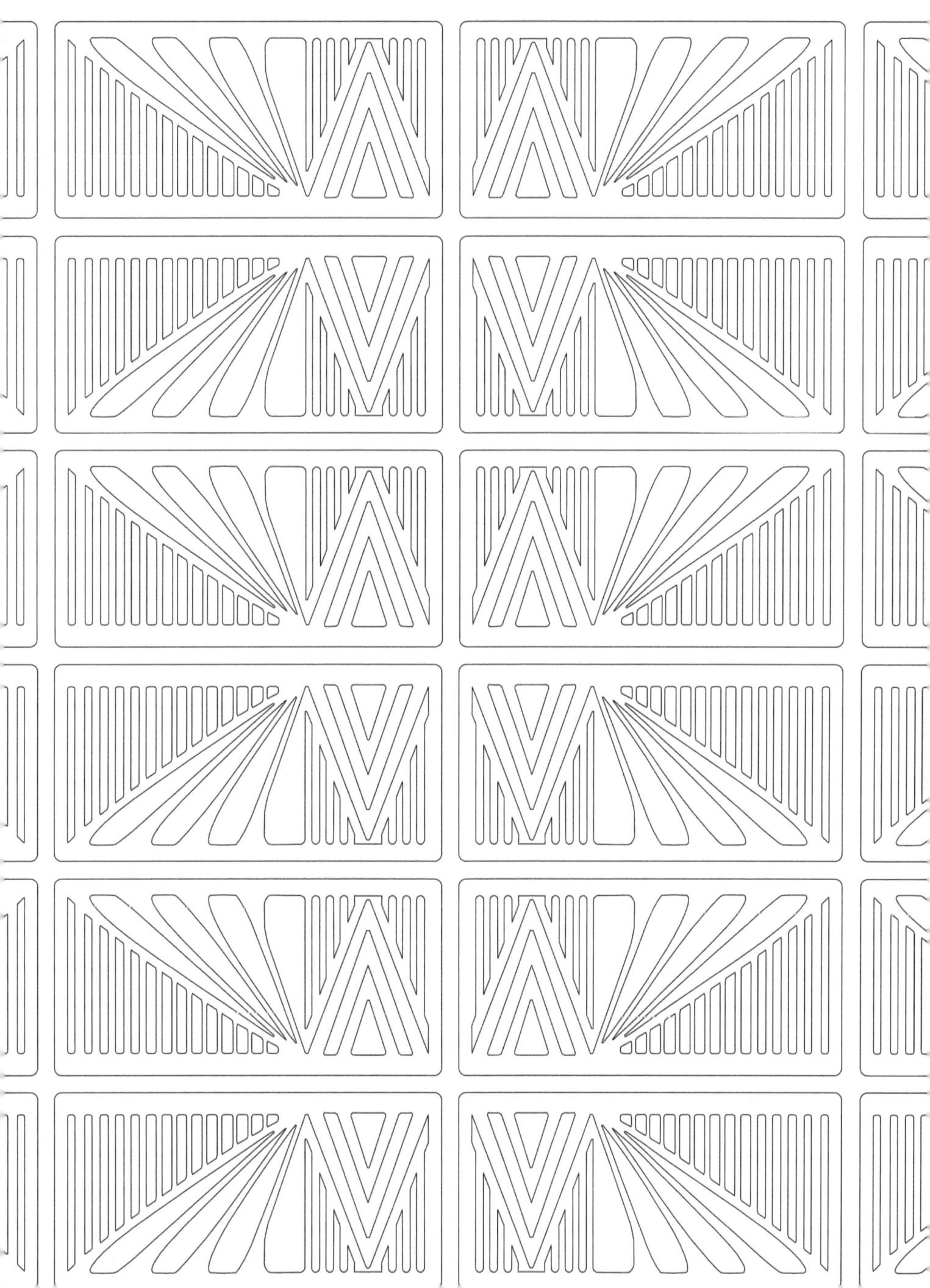

State Theatre Carpet

This wonderfully complex Art Deco carpet design is a replica of the original carpet design. The pattern was recreated during the 2017 restoration based on a sample that a local historian had preserved. The carpet is composed of five different colors, featuring rose, green, and orange tones.

**STATE
THEATRE**
ANN ARBOR

State Theatre Lobby

The Mezzanine Lounge of Ann Arbor's State Theatre is a truly lovely Art Deco design by the renowned C. Howard Crane. The swirling look of the kidney bean-shaped recessed ceiling, the large round structural columns with elegant wood veneer, the built-in curved and linear banquette seating, and the interesting and abstract shapes of the unique and colorful carpeting, creates an elegant access point to the theater's second-floor essentials.

Originally, and before there was an elevator, this area provided access to balcony seating and the womens and mens' lounges. The ladies lounge featured a high-style Art Deco "Powder Room" with elegant art moderne-style furniture and painted decorations on the walls. The men's lounge featured a "Smoking Room" with built-in high-style deco benches and stylish moderne tile on the walls.

Today, the Mezzanine Lounge can be accessed, as can all four of the theaters, by an elevator to all levels, plus an escalator will whisk you from the Mezzanine Lounge to theaters three and four. This wonderful Art Deco space was restored to the original color palette, as were the banquette seats and the round wood veneer columns. Added to the Mezzanine Lounge is a modern concession stand, which feels as if it has always been there (but was added in 2017) and a restored, 1940s-era Wurlitzer 78rpm Juke Box.

STATE THEATRE ANN ARBOR

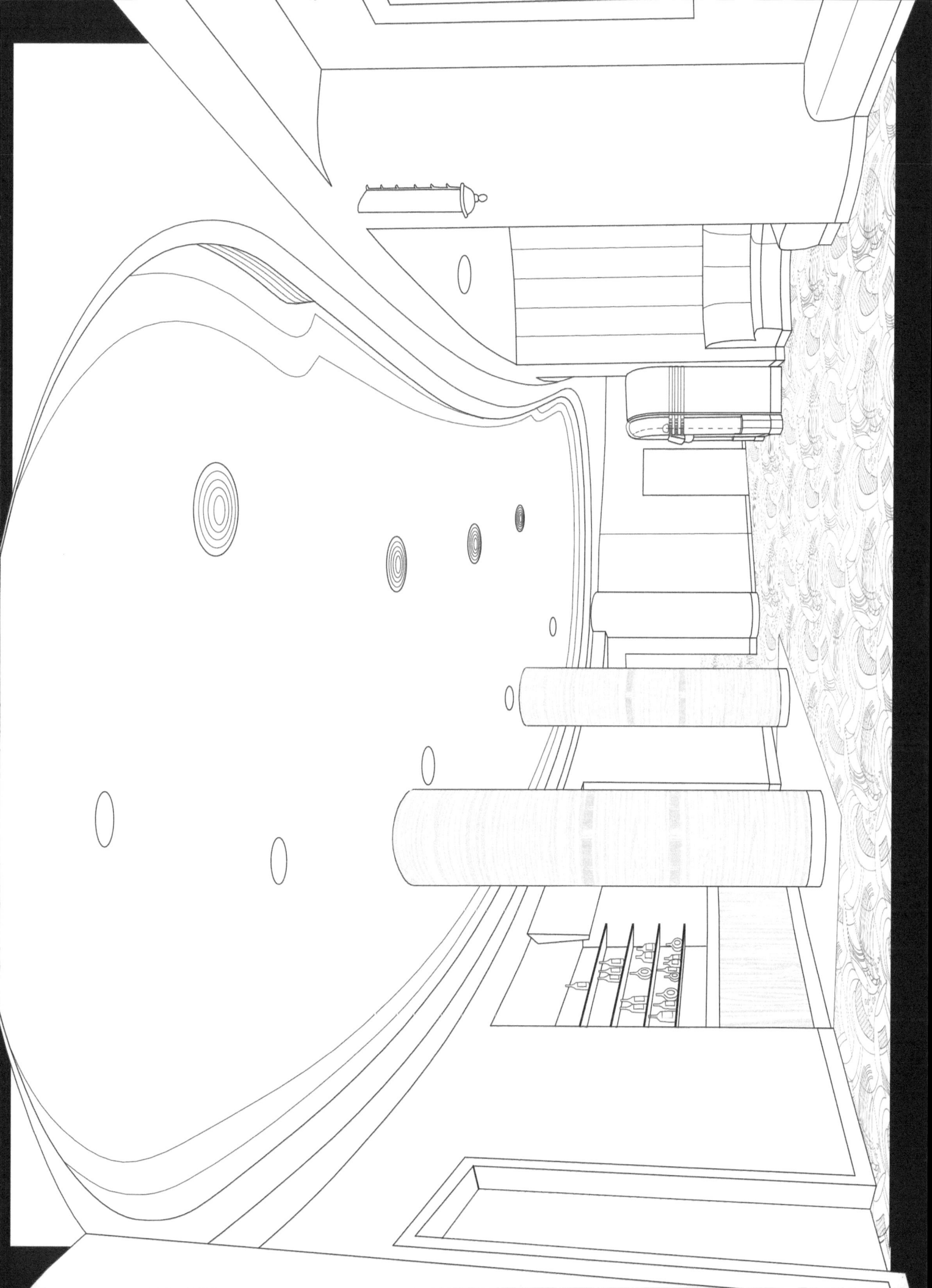

State Theatre Auditoriums

The auditoriums have been renovated to feature a world-class viewing experience with Art Deco elegance.

Each of the four auditoriums features stadium seating with rich fabric in green, orange, and gold tones. The four theaters are slightly different in their accents and patterns, but are all stunning examples of Art Deco design.

A favorite feature of the State Theatre auditoriums are the majestic sconces placed alongside each screen. Beautifully crafted and fully functional, these historically sensitive replicas help make the viewing experience in any of the State Theatre auditoria truly magical.

STATE THEATRE ANN ARBOR

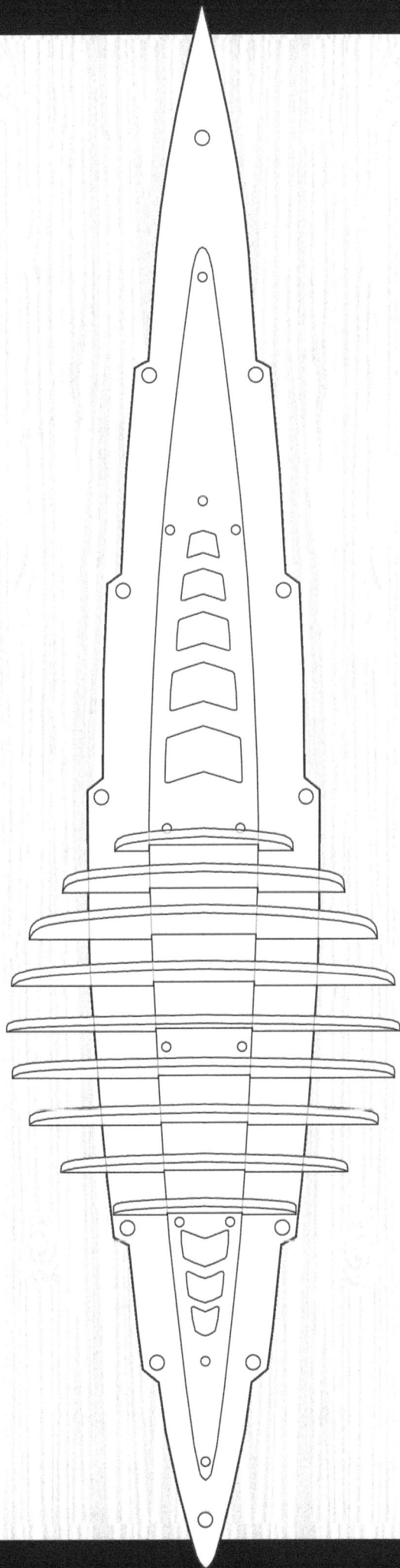

About the Author

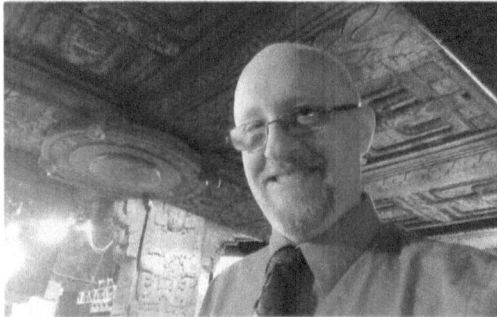

Escott O. Norton grew up in the worlds of both theater and architecture. His mother Sally co-founded the Occidental College Summer Drama Festival and his father, Oakley, a math teacher and progressive education pioneer, designed and built all of Escott's childhood homes.

From an early age, Escott was drawn to the historic theatres of Los Angeles. As a young kid his Mom frequently took him to the Eagle and the Rialto, and as a pre-teen he took the bus to Hollywood to see movies at Grauman's Chinese, the Cinerama Dome, and other theatres. He founded Friends of the Rialto in 1983 to advocate for its protection, and was a charter member of the Los Angeles Historic Theatre Foundation, an organization he was proud to lead as executive director for six years.

Escott has traveled all over the world with his amazing wife Jeanne and sons Kyle and Evan, and is grateful for their constant support and patience as he insisted on stopping at every historic theatre he could find!

He continues to advocate for the preservation and revitalization of historic theatres, and his company, EON Design Co, does design consultation for historic theatres, commercial buildings, and residences.

For more information:
EONDesignCo.com   ColoringHistoricTheatres.com

Special Thanks

The author gratefully acknowledges the help of the following people who were crucial to the creation of this book series: Jeanne S. Norton, Natalie Norton, and Bill Counter

—⋅◦∞◦⋅—

Helpful Resources

The original illustrations and text in this book are based on information collected by the author and supplied by Marquee Arts. For more information on historic theatres and related subjects, here are some great resources:

Mike Hume's Historic Theatre Photos
HistoricTheatrePhotos.com

Bill Counter's Los Angeles Theatres Blog
LosAngelesTheatres.blogspot.com

Theatre Historical Society
HistoricTheatres.org

Steve Simon's Fanchon and Marco Page
FanchonandMarco.com

These preservation advocacy groups are doing great work to preserve and protect historic theatres in Los Angeles and across the country. Please support them by becoming a member!

Los Angeles Historic Theatre Foundation
LAHTF.org

League of Historic American Theatres
LHAT.org

Many areas have regional preservation groups that work to protect historic buildings, including theatres. Check out and support your local preservation group. One of the largest groups in the country is located in Los Angeles:

Los Angeles Conservancy
LAConservancy.org